AF270476

Chasing
BIGFOOT

Anna Anderhagen

Big Buddy Books

An Imprint of Abdo Publishing

abdobooks.com

abdobooks.com

Published by Abdo Publishing, a division of ABDO, PO Box 398166, Minneapolis, Minnesota 55439. Copyright © 2024 by Abdo Consulting Group, Inc. International copyrights reserved in all countries. No part of this book may be reproduced in any form without written permission from the publisher. Big Buddy Books™ is a trademark and logo of Abdo Publishing.

Printed in the United States of America, North Mankato, Minnesota
102023
012024

Design: Denise Hamernik, Mighty Media, Inc.
Production: Mighty Media, Inc.
Editor: Liz Salzmann
Cover Photograph: altitudevisual/Adobe Stock
Interior Photographs: Adam Jones/Flickr, p. 11; Bettmann/Getty Images, p. 13; bonciutoma/Adobe Stock, p. 29; Chuck Stoody/AP Images, p. 27; Frederick M. Brown/Getty Images, p. 21; LLOYD YOUNG/AP Images, p. 25; MICHAEL MACOR/AP Images, p. 19; Petrarch1603/Wikimedia Commons, p. 15; Raggedstone/Shutterstock Images, p. 5; Victoria/Adobe Stock, p. 7; Worldillustrator/Adobe Stock, p. 9; Zack Frank/Adobe Stock, p. 17
Design Elements: Adobe Stock (silhouette); Joko/Adobe Stock (tape and paper); Melica/Adobe Stock (polaroid frame); Net Vector/Shutterstock Images (map); Oleg Iatsun/Shutterstock Images (compass); Piman Khrutmuang/Adobe Stock (paper); SlipFloat/Adobe Stock (footprints); STILLFX/Shutterstock Images (background texture); Wandeaw/Shutterstock Images (background texture); Yevhenii/Adobe Stock (tape)

Library of Congress Control Number: 2023939280

Publisher's Cataloging-in-Publication Data
Names: Anderhagen, Anna, author.
Title: Chasing bigfoot / by Anna Anderhagen
Description: Minneapolis, Minnesota : Abdo Publishing, 2024 | Series: Chasing cryptids | Includes online resources and index.
Identifiers: ISBN 9781098291884 (lib. bdg.) | ISBN 9781098278786 (ebook)
Subjects: LCSH: Sasquatch--Juvenile literature. | Yeti--Juvenile literature. | Animals, Mythical--Juvenile literature. | Folklore--Juvenile literature. | Cryptozoology--Juvenile literature.
Classification: DDC 001.944--dc23

CONTENTS

CHAPTER 1

CHASING CRYPTIDS

Ew, what is that smell? You and your friend stop on the forest path. A very tall, hairy creature just hid behind a tree! It looks part human and part animal. Could this be the Bigfoot cryptid that you have been looking for?

FAST FACT

Bigfoot **expert** Michael Rugg says Bigfoot smells like "a skunk that rolled around in dead animals and hung around **garbage** pits."

Many hikers have reported seeing Bigfoot in forests throughout North America.

CHAPTER 2

WHAT IS A CRYPTID?

A cryptid is an animal that has not been proven to exist. There are stories about many different cryptids. One of the best-known cryptids is Bigfoot.

Many people think they have seen Bigfoot, but no one has been able to prove it. Most people **describe** Bigfoot as at least 7 feet (2 m) tall, covered in hair, and very smelly.

Areas with many Bigfoot sightings often have Bigfoot-themed shops.

CHAPTER 3

FIRST RECORDED SIGHTINGS

Native Americans were the first people to talk about Bigfoot. They called it Sasquatch.

In the mid-1800s, pioneer Elkanah Walker met with Native Americans in Oregon. He wrote down their stories of giants in the mountains. The Native Americans said the giants stole their fish while they slept.

CRYPTID PROFILE

NAMES: Bigfoot, Sasquatch

CLASSIFICATION: mammal

COUNTRIES: Canada, United States

HABITATS: forests, mountains, creeks, swamps, rivers

DESCRIPTION:
  — muscular
  — walks on two feet
  — 6 to 15 feet (1.8 to 4.6 m) tall
  — covered in black, dark brown, red, or gray hair

PROVEN TO EXIST: not yet

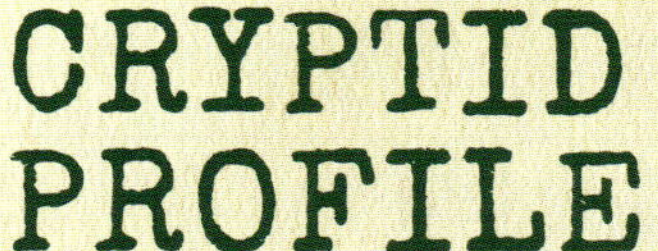

CHAPTER 4

HIKING IN THE MOUNTAINS

In 1955, William Roe was hiking in Canada. He saw a tall, hairy creature with wide shoulders. It had long arms that went down to its knees.

At first, Roe thought it was a bear. He almost shot it with his rifle. But when the creature looked at him, it looked part human. Roe lowered his gun and watched it walk back into the forest.

Roe spotted Bigfoot in the forest on Mica Mountain in British Columbia, Canada.

CHAPTER 5

CAUGHT ON FILM?

In 1967, Roger Patterson and Bob Gimlin filmed a hairy creature walking by a creek. They said it had to be Bigfoot. Many people think the film is fake. Some say the creature was a person in a gorilla costume. But others argued that the creature's movements and body didn't look human.

An image of Bigfoot from the film Patterson
and Gimlin made. Do you think the film is fake?
Why or why not?

CHAPTER 6

MORE SIGHTINGS

In 2001, a hiker said he saw Bigfoot fishing in a creek near Dunsmuir, California. The hiker said the creature was about 7.5 feet (2.3 m) tall and had very dark long hair.

In 2005, a backpacker on Silver Star Mountain in Washington State saw something strange. He thought it was a large rock at first. He took some pictures of it. Then the rock started moving!

Silver Star Mountain is in Gifford Pinchot
National Forest. The mountain has several
hiking trails.

In 2023, Bigfoot was spotted in Heflin, Alabama. A husband and wife said they heard strange sounds in the woods near their house for two nights. On the third night, they looked into the woods and saw Bigfoot peeking around a tree.

Heflin is located in Talladega National Forest near the edge of the Appalachian Mountains.

TRACKING BIGFOOT

There are groups that track Bigfoot. These groups look for footprints and other clues in the wilderness. Some use **drones** to try to find Bigfoot.

Bigfoot trackers also record sounds that could be made by Bigfoot. Some Bigfoot **researchers** have recorded grunts, growls, screams, and howls. But no one has been able to prove that these sounds are from Bigfoot.

Bigfoot hunter John Freitas looks for Bigfoot in California. He blasts recorded Bigfoot calls from his truck. He hopes to record a response from Bigfoot.

CHAPTER 8

BIGFOOT ORGANIZATIONS

The Bigfoot Field **Researchers** Organization (BFRO) **investigates** reports of Bigfoot sightings. It enters the ones it thinks are **credible** into the BFRO database.

The Bigfoot Mapping Project created an online map of Bigfoot sightings. Anyone who sees Bigfoot can add their sighting to the map.

Matt Moneymaker (*far right*) started the BFRO. In 2012, he and other Bigfoot experts spoke at a conference in California.

BIGFOOT SIGHTINGS

Reported Bigfoot sightings since 2020, according to the BFRO

⊙ = one sighting

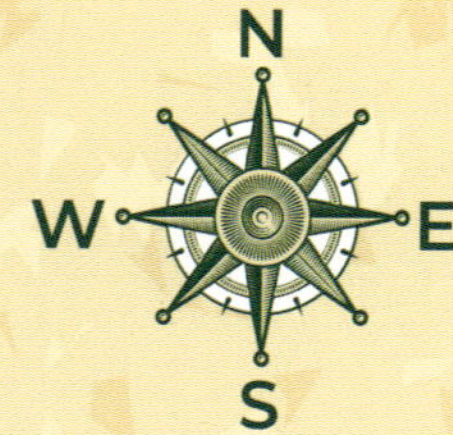

CANADA

STUDYING BIGFOOT

Some scientists think Bigfoot could be a form of the *Gigantopithecus*. This was the largest known ape. It lived more than 100,000 years ago. Scientists study **fossils** of its bones to see if there is a Bigfoot connection.

Scientists also test hair found near Bigfoot sightings or footprints. The hair tested so far has come from animals such as bears, deer, or wolves.

Illinois State University professor Angelo Capparella studies a hair sample that was found near a Bigfoot sighting.

CHAPTER 10

CRYPTID KEEPER: GROVER KRANTZ

Grover Krantz was a **cryptozoologist** and **anthropologist**. Krantz believed Bigfoot existed. He spent years trying to prove it.

Krantz talked to people who said they saw Bigfoot. Then he **examined** clues found in areas near sightings. But Krantz died in 2002 without having found the proof he needed.

Krantz made many casts of footprints that he thought could be Bigfoot's.

CHAPTER 11

DOES BIGFOOT EXIST?

Many people think they have seen Bigfoot. There are hundreds of drawings, footprint sightings, blurry photographs, and videos of Bigfoot. But no one can prove they are real.

Most scientists do not think Bigfoot exists. They suggest that people mistake bears or deer for Bigfoot. What do you think?

If you were walking in the forest, would you
want to see Bigfoot?

anthropologist—a person who studies the beginnings, development, and behaviors of humans and their ancestors.

credible—believable.

cryptozoologist—someone who looks for and studies legendary animals to determine whether they exist.

describe—to tell about something with words. Such a telling is a description.

drone—an aircraft or ship that is controlled by radio signals.

examine—to look at closely.

expert—a person with special skills or knowledge on a subject.

fossil—a trace of an organism from a past geologic age that has been preserved in the earth's crust.

garbage—trash.

investigate—to gather information about or study something.

researcher—a person who carefully studies a subject in order to learn facts about it.

ONLINE RESOURCES

To learn more about Bigfoot, please visit **abdobooklinks.com** or scan this QR code. These links are routinely monitored and updated to provide the most current information available.

INDEX